Little Star ...
Raising Our First Calf

By Twins Rianna and Sheridan Chaney

Published by Down Under Publications

Second Edition, March 2011
Copyright, Rianna and Sheridan Chaney, 2008
ISBN 978-0-9818468-0-4

(301) 271-2732
Chaneyswalkabout@aol.com

Edited by Rebecca Long Chaney
Photographs by Kelly Hahn Johnson and Rebecca Long Chaney

Layout and Design by Kathy Moser Stowers

To our Daddy and Mommy,

for giving us the responsibility of raising our very first farm animal and for sharing all the joys of growing up on a farm.

We would like to thank Elizabeth Randall (Miss Betsy) and her late husband George Randall for buying our Pappy and Meme's (Harold and Peggy Long) farm and for asking our Daddy to manage the land and the registered Herefords that Mr. George so loved. We are so proud to help Miss Betsy with these beautiful animals and know that Mr. George would be very happy.

And a huge thank-you to our mommy's friends, Kelly Hahn Johnson and Kathy Moser Stowers, for helping get our first children's book completed. We also thank the many friends, relatives and experts who looked at our book and offered their advice, support and encouragement. And a big thank-you to Jolene Brown who gave Mommy the idea for this project.

We were three years old when one fall evening Daddy took us to see the newborn calves. They were twins just like us. They were so small and cute, but their mother, the cow, could only take care of one of them. That made us sad. When Daddy said we could take care of the other calf, we were so excited!

We named her, Little Star. She is a heifer, which means she is a female. We are feeding Star a milk replacement made of powdered milk and special vitamins. We mix it with warm water in a pitcher using a whisk from our mommy's kitchen. The vitamins help keep Star healthy.

It's not easy mixing and pouring the milk into the bottle. We always work together. Oops!!! Our Australian Shepherd dog, Bode, loves it when we spill milk because he gets to lick it all up.

Star can't wait to get her bottle every morning and every evening. The bottle and nipple look just like our baby bottles did when we were little, but Star's bottle is big to make sure she gets lots of milk to drink. Look at that nipple, it's giant!

We love to watch Star jump around and kick up her hooves. Her milk must be yummy by the way she licks her nose. Mommy says we always have to wear our barn boots because you never know what you might step in — Watch out for that cow poop, Sissy!

Star is a Hereford beef calf and has red and white fluffy hair. She will be raised to keep at our farm and have her own calves one day, our Daddy told us. When it's feeding time, we just call Star's name and she comes running. We stand on the gate and wait for her sometimes. We try to pet her, but she tries to lick our hand.

Sometimes we both have to hold onto the milk bottle because Star is a strong sucker and pushes it. Star wears a big yellow earring that Daddy calls an ear tag. It tells Daddy who Star's parents are and when Star was born.

If Star's mother could have taken care of her, Star would have nursed her mother's udder just like this calf is doing in our field. A beef calf usually nurses its mother for six months before it is weaned. We don't think Star misses her mother because we give her so much love.

See how funny our Mommy looks! She thought she would put a halter on Star so we could train her to lead. Star likes to come to us to drink her bottle of milk but didn't want to have anything to do with Mommy and that halter.

Mommy is pushing, and Sissy is pulling, but Star is being stubborn. Mommy is always right beside us when we are teaching Star to walk with a halter on. If we can train Star to lead, we can take her to the fair to win a prize.

Cousin David is here almost every day and enjoys helping us with Star. Our friend Brady, from town, comes to visit and he gets to learn about farm animals. He learns quickly that leading Star is not an easy job.

Sissy, looks like Little Star thinks you taste good. Many farm animals become calm because the owners give them special care every day. Farmers and visitors should still be very careful around all farm animals and children should always have adults with them.

Star is growing so big and we are still working as a team. The bucket with Star's grain is heavy so we both carry it. Our farm chores get done quicker when we work together. We have learned a lot about raising a calf.

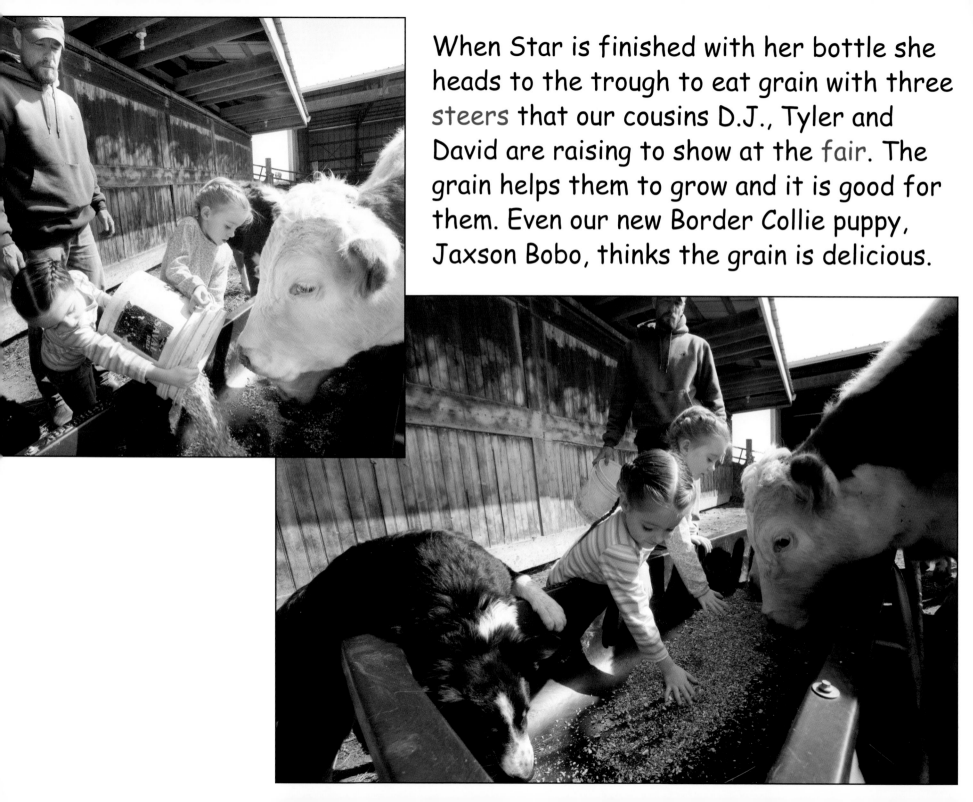

When Star is finished with her bottle she heads to the trough to eat grain with three steers that our cousins D.J., Tyler and David are raising to show at the fair. The grain helps them to grow and it is good for them. Even our new Border Collie puppy, Jaxson Bobo, thinks the grain is delicious.

Star likes when we throw hay in the hay rack. During the summer our Daddy rolls dried grass called hay into big round bales. The animals eat it all winter when there's not much grass in the field.

Mommy says it's important to learn about taking good care of our farm animals. We are learning a big word, responsibility, which means doing the little chores we are asked to do. Jaxson Bobo loves to stick his head the whole way in the pitcher to lick up the leftover milk.

We like to go with Daddy to check the cows to see if there are any new calves. February is here and there are 58 cows to have calves. We love to see the calves run and play just like our Star likes to do.

We give Star tiny alfalfa cubes. They are just like dog treats but for calves and cows. Alfalfa cubes are dried hay packed together. We can't wait until Daddy gives us another little calf to raise. It was so much fun, but a lot of work, too!

Star gets prettier every day and she likes when we brush her. We use special cow brushes to smooth Star's soft hair. When it warms up we are going to give Star a bath using a hose and bucket.

Keeping the barnyard clean is important because it is where Star spends a lot of her time. We help Mommy and Daddy with chores and sometimes shovel the cow poop. We push it to the big manure pile for Daddy to take it away.

There have been 31 calves born in the past three weeks. We like to see the calves nap in the warm sun and play together. Sometimes we take a walk to look at the calves to make sure they are healthy and the cows are taking care of them.

Star is very special to us. Star will soon be weaned off the bottle and Daddy will put her with all the other calves on pasture to enjoy the fresh spring grass. We sure will miss giving Little Star hugs!

This is our favorite time of day, when the sun is setting over the mountains where we live. God makes the colors so beautiful. It's always fun to take walks around the farm with Daddy and Mommy.

Glossary

Barnyard - The area near the barn where farm animals often live.

Farm - A large area that may grow crops and raise livestock.

Fair - An event focusing on farm animals, the good care they get and contests for their confirmation as well as how they lead.

Heifer - A female calf.

Hereford - One of several breeds of beef cattle.

Hooves - The hard part of the calf's toe or foot.

Manure - A fancy name for cow poop.

Milk Replacement - A powdered milk made of special vitamins.

Pasture - Areas of grass for farm animals to eat.

Steer - A bull calf that has been neutered.

Udder - The milk bag with four teats under a cow where beef calves usually get all their nourishment the first few months.

Wean - The time when calves are taken off milk and are able to digest grass, grain and hay and continue to be healthy animals.

Ag Facts

Star is a Hereford beef calf. A few of Mommy's other favorite beef breeds include Angus, Limousin, Charolais, Braunvieh, Shorthorn, Brahman, Simmental, Devon, Texas Longhorn and Scottish Highland cattle.

American farmers are the most productive farmers in the world. Today, 98% of all U.S. farms are owned by individuals, family partnerships or family corporations. Each United States farmer produces enough food and fiber for 143 people in the United States and around the world.

Born two months premature, Rianna and Sheridan Chaney have always been fighters. They enjoy the outdoors, and before they could even walk would enjoy rides through the field on the six-wheeled Gator farm vehicle. They sat on their parents' laps to take in the sights, sounds and "smells" of farm life.

They love to get dirty, play outside, dress up like princesses and enjoy time with their grandparents, aunts, uncles and cousins. Born two hours apart the girls remain inseparable and call each other Sissy.

Rebecca Long Chaney is a full-time m[...] to four-year-old twin daughters Ria[...] and Sheridan. She is an author, awa[...] winning freelance journalist and speak[...] She has shared her inspirational messa[...] "Dare to Risk Life Change" w[...] audiences across the country. T[...] keynote address is based on the li[...] changing adventure "before childr[...] with her husband, Lee, that led to [...] publication of "Bulldust In My Bra[...] An American Couple's Working Season in the Outback." T[...] Chaneys have been back in Maryland for six years living on [...] old family farm that had been in the Long family more than [...] years before the Randalls purchased it in December, 2001. I[...] manages Randall Land and Cattle Company, a registe[...] Hereford beef operation.

For more information about Rebecca's speaking availabi[...] or the twins' next children's book, contact Rebecca at (301) 2[...] 2732, or e-mail her at chaneyswalkabout@aol.cc[...] info@rebeccalongchaney.com or check out her Web site[...] www.rebeccalongchaney.com.

Star was born September 22, 2007. She has a twin sister that was raised by her mother and Star was raised by Rianna and Sheridan. Star will live out her days at Randall Land and Cattle Company in the breeding herd and as a very special animal to the girls.

Award-winning photographer Kelly Hahn Johnson not only is known for her unique photojournalism style but is known for her approach to portraiture. She's won countless awards and her images have been featured in local, state and national publications. She loves observing people, moments and emotions, creating images that will be treasured for years to come, like the priceless images she's captured of the Chaney twins over the years. She lives with her husband Bl[...] and son Brady in a century-old renovated house in Sharpsb[...] Maryland.

Visit her online gallery at kellyhahnphotography.com. Con[...] Kelly at info@kellyhahnphotography.com or call her at 240-2[...] 3677.

Kathy Moser Stowers lives with her husband and children on her family's farm near Jefferson, Maryland. She has 22 years of experience in design and layout at *The Frederick News-Post*. She has shared her talents and expertise with local, state, national and international groups, organizations and events. She specializes in brochures, flyers and business cards.

For help in designing brochures, pamphlets, books, business cards, etc., contact Kathy at Wakstowers@aol.com or call 301-371-9306.